Marine Chemistry

A report of the Marine Chemistry Panel of the Committee on Oceanography

NATIONAL ACADEMY OF SCIENCES Washington, D.C. 1971

NOTICE: The study reported herein was undertaken under the aegis of the National Research Council's Committee on Oceanography. The institutional responsibilities of the National Research Council were then discharged in the following manner:

The members of the study committee were selected for their individual scholarly competence and judgment with due consideration for the balance and breadth of disciplines. Responsibility for all aspects of this report rests with the study committee, to whom sincere appreciation is expressed.

Although the reports of study committees are not submitted for approval to the Academy membership nor to the Council, each report is reviewed by a second group of scientists according to procedures established and monitored by the Academy's Report Review Committee. Such reviews are intended to determine, *inter alia*, whether the major questions and relevant points of view have been addressed and whether the reported findings, conclusions, and recommendations arose from the available data and information. Distribution of the report is permitted only after satisfactory completion of this review process.

ISBN 0-309-01928-1

Library of Congress Catalog Card Number 71-177418

Available from

Printing and Publishing Office
National Academy of Sciences
2101 Constitution Avenue, N.W.
Washington, D.C. 20418

Printed in the United States of America

The cover design is from a 1,200-km-long, 1,000-m-deep oceanographic section perpendicular to the Peruvian coast at 17°S in November of 1960. Of special interest is the subsurface core of water of extremely low (less than 0.2 ml per liter) dissolved oxygen content flowing southward here in the Peru–Chile Undercurrent.

Dedication

The members of the Committee responsible for this report dedicate it to their chairman, Norris W. Rakestraw, the dean of the marine chemists of the United States.

Norris was one of the mere handful of investigators who forty years ago were seeking an understanding of the composition of seawater. He has witnessed an explosive growth in the number of practitioners of his vocation, many of whom have benefited from stimulation and guidance by him. He has seen seawater chemistry prosper through the development of techniques of increasing precision and sensitivity and has seen it applied with great profit in unraveling problems in geochemistry, paleoclimatology, and marine biology.

Norris is the personification of the scholarly teacher. His scientific career has been spiced by travels about the world in which he takes contagious delight. His personal fulfillment in gaining better understanding of the characteristics of our environment is an attribute that has won him recognition by fellow chemists and educators.

Norris has never been content with what we *think* we know. He has concerned himself with what we know we *don't know*. Hence, we have this book of questions. Because the subject matter is too vast for one man to handle alone, he has drawn on nine of his colleagues to consider the present state of the art in marine chemistry and to define some of the priorities for research in coming years.

Although Norris has left the academic center of oceanography at La Jolla for the dryer climate of a ranch at Palm Springs, this volume shows that he has not abandoned his vocation. Through it may his influence on marine science continue.

ALFRED C. REDFIELD

Woods Hole, Massachusetts
June 14, 1971

Preface

In the spring of 1968, the National Academy of Sciences Committee on Oceanography organized a panel on Marine Chemistry under the chairmanship of Professor Norris A. Rakestraw. The Panel met in October of that year, and a proposed report on marine chemistry was outlined and writing assignments made to panel members. During the fall and winter of 1968, letters were sent by Professor Rakestraw to some 80 scientists working in the field of marine chemistry inviting their suggestions on a number of topics, including the scope of marine chemistry, the major problems in the field, advances that have been made, and future trends and needs.

Much of the work on the report during the next several months was done by correspondence. The Panel met again in July of 1969 at the Gordon Conference on Chemical Oceanography at Meriden, New Hampshire. More writing and correspondence among Panel members produced a rough draft of most of the report by the spring of 1970. Panel members Goldberg, Garrels, and Rakestraw collated and digested this material, and the final manuscript was completed in August of 1970.

The report was subsequently reviewed by the Committee on Oceanography and the Division of Earth Sciences of the National Research Council.

Preparation of this report was made possible by the support provided to the Ocean Affairs Board (successor to the Committee on Oceanography in December 1970), by the Office of Naval Research, the Atomic Energy Commission, the National Science Foundation, the National Oceanic and Atmospheric Administration (previously the Interior Department and Environmental Science Services Administration) and the U.S. Coast Guard.

RICHARD C. VETTER
Executive Secretary
Ocean Affairs Board

Marine Chemistry Panel

NORRIS RAKESTRAW, *Chairman*, Morongo Valley, California
RICHARD BADER, Rosenstiel School of Marine and Atmospheric Sciences, University of Miami
JOHN BUNT, Rosenstiel School of Marine and Atmospheric Sciences, University of Miami
JAMES CARPENTER, Head, Oceanography Section, National Science Foundation
DAYTON CARRITT, Director, Institute for Man and His Environment, University of Massachusetts
GORDON ERDMAN, Phillips Petroleum Company Research Center, Bartlesville, Oklahoma
ROBERT M. GARRELS, Scripps Institution of Oceanography
EDWARD GOLDBERG, Scripps Institution of Oceanography
JOHN HUNT, Woods Hole Oceanographic Institution
DAVID MENZEL, Skidaway Institute of Oceanography
TIMOTHY PARSONS, Institute of Oceanography, University of British Columbia
RICARDO M. PYTKOWICZ, Department of Chemical Oceanography, Oregon State University

Contents

1

Introduction

In this report, we describe the field of marine chemistry, the activities of marine chemists, and the nature of the principal problems. In a very broad way, we also indicate the directions in which we may expect progress to be made and various ways in which that progress may be hastened. We also consider the supply of and demand for workers in the field and the kind of training they should have.

This is not a textbook on marine chemistry. It is not intended for the expert, who will be unlikely to find anything in it that is new to him. It is directed to the interested public and, more particularly, to workers in neighboring scientific fields and to those in other subdivisions of oceanography who may wish to know how the work of marine chemists is related to their own interests.

The reader will see that chemistry not only contributes to oceanography but that it is essentially involved in the attack on some of the most important problems in the study of the ocean. Some older workers in the field have been of the mind that chemistry performed upon seawater is not an end in itself but a service to other disciplines. Their mission has been to apply the principles and methods of chemistry to the solution of problems in oceanography. They have, accordingly, long been known as chemical oceanographers. However, it has been argued that they should be generalists with backgrounds and interests in physics, biology, and geology, fields with which their own is closely related in the study of the ocean.

Although the mission of the older chemical oceanographers is still appropriate as far as it goes, it is being modified and supplemented, particularly by the advent of new workers, drawn largely from the general field of chemistry, who bring with them new ideas and the conception of new problems. The substitution of "marine chemistry" for "chemical oceanography" is evidence of this new approach. For example, it is now being appreciated that the ocean is an environment in which chemistry can be studied for its own sake. There is no other aqueous system in which so many components have existed simultaneously for so long a time.

A changing view of the function of marine chemistry has accompanied a change in the public conception of "oceanography." Whereas professionals have long defined oceanography as the scientific study of the ocean and its behavior, distinct from the exploitation of its resources, the latter has been emphasized in the movement to gain wider public appreciation and support, and now the word "oceanography" connotes virtually everything remotely connected with the sea.

However, this report is not concerned with the chemistry of petroleum (now widely considered a marine resource), the composition of antifouling paints (used on the bottom of ships), or other areas of chemistry related to ocean technology. We do discuss chemical phenomena that are directly concerned with the ocean as such. In doing so we discuss not only the water but also that portion of the ocean bed that settled from the water, as well as the indefinable layer of the atmosphere with which the ocean is in constant exchange. We have neglected the basaltic layer of the earth's crust that is under the ocean, and we have considered that rivers are not extensions of the ocean but rather sources of its material.

Delineation of the environment is still far from complete. Estuaries, for example, are probably extensions of the ocean, but we cannot stipulate how low the salinity must become before the water ceases to be marine. We must certainly deal with the liquid water of the ocean, but should we include the particulate matter suspended in it? Some detrital particulate matter takes an active part in ocean chemistry processes, and living forms that have originated and spent their lives in the sea certainly have a claim to membership in the marine chemical environment. On the other hand, floating garbage, logs, and steamships, though obviously in the ocean, are foreign and their chemical composition is outside our consideration. As a guideline, we have defined the ocean to include—but not to be entirely limited to—the water and everything that has originated in it. There must be a valid reason for further inclusions. Sediments are a case in point. Much of this material has been washed into the ocean from the land, but this terrigenous material interacts with the seawater and cannot be neglected.

Marine biology presents a special problem. It has always been difficult to distinguish marine biology from biological oceanography. If we regard an organism as an integral part of the ocean simply because it happens to live there (and we must therefore consider its chemical composition and metabolic reactions), we risk involving ourselves in a needless repetition of general biochemistry. Although our criteria are flexible, we have focused on those aspects and factors that are to some extent characteristic of the marine environment or important to the chemistry of the ocean.

When an elementary student makes his first contact with chemistry, he learns that it is the science that deals with the properties and behavior of the different kinds of matter and with the conditions under which one kind of matter can be changed into another. However, of the hundreds of thousands of different kinds of matter that have been identified, only a relatively small proportion occur in nature. Geochemistry is the science that deals with these naturally occurring substances, their sources and distribution, and all that happens to them during natural processes. Marine chemistry can be considered as a special branch of geochemistry with the added complication of the ocean biota.

Marine chemistry starts with the problem of identifying and describing the different kinds of matter in the marine environment and the transformations that take place among them, i.e., the chemical processes occurring in the ocean.

Table 1 lists the various elements that have been identified in the waters of the ocean and their concentrations, as well as we know them at the present time. Some of the concentrations are highly variable and the values given are merely representative. Eighty-four of the basic elements are included. It is very likely that all elements are present in the ocean but have not yet been detected.

There are several viewpoints from which the chemist can approach the ocean. He can consider the ocean as

TABLE 1 Composition of Seawater

Element	Seawater Concentration (μg/liter)	Principal Dissolved Species
H	1.1×10^8	H_2O
He	7×10^{-3}	He (gas)
Li	1.7×10^2	Li^+
Be	6×10^{-4}	–
B	4.5×10^3	$B(OH)_3$, $B(OH)_4^-$
C	2.8×10^4	HCO_3^-, CO_3^{-2}
C (organic)	2×10^2	–
N	1.5×10^4	N_2 (gas)
N	6.7×10^2	NO_3
O	8.8×10^8	H_2O
O	6×10^3	O_2
O	1.8×10^6	SO_4^{-2}
F	1.3×10^3	F^-
Ne	0.12	Ne (gas)
Na	1.1×10^7	Na^+
Mg	1.3×10^6	Mg^{+2}
Al	1	–
Si	3×10^3	$Si(OH)_4$, $SiO(OH)_3^-$
P	90	HPO_4^{-2} $H_2PO_4^-$, PO_4^{-3}
S	9.0×10^5	SO_4^{-2}
Cl	1.9×10^7	Cl^-
Ar	4.5×10^2	Ar (gas)
K	3.9×10^5	K^+
Ca	4.1×10^5	Ca^{+2}
Sc	$< 4 \times 10^{-3}$	$Sc(OH)_3$
Ti	1	$Ti(OH)_4$
V	2	$VO_2(OH)_3^{-2}$
Cr	0.5	CrO_4, Cr^{+3}
Mn	2	Mn^{+2}
Fe	3	–
Co	0.4	Co^{+2}
Ni	7	Ni^{+2}
Cu	3	Cu^{+2}
Zn	10	Zn^{+2}
Ga	3×10^{-2}	–
Ge	7×10^{-2}	$Ge(OH)_4$
As	2.6	$HAsO_4^{-2}$, $H_2AsO_4^-$
Se	9×10^{-2}	SeO_4^{-2}
Br	6.7×10^4	Br^-
Kr	0.2	Kr (gas)
Rb	1.2×10^2	Rb^+
Sr	8×10^3	Sr^{+2}
Y	1×10^{-3}	$Y(OH)_3$
Zr	3×10^{-2}	–
Nb	0.01	–
Mo	10	MoO_4^{-2}
Rh	7 to 11×10^{-3}	–
Ag	0.3	$AgCl_2^-$
Cd	0.1	Cd^{+2}

TABLE 1 (continued)

Element	Seawater Concentration (μg/liter)	Principal Dissolved Species
In	4×10^{-3}	–
Sn	0.8	–
Sb	0.3	–
I	60	IO_3^-, I^-
Xe	5×10^{-2}	Xe (gas)
Cs	0.3	Cs^+
Ba	20	Ba^{+2}
La	3×10^{-3}	$La(OH)_3$
Ce	1×10^{-3}	$Ce(OH)_3$
Pr	0.6×10^{-3}	$Pr(OH)_3$
Nd	3×10^{-3}	$Nd(OH)_3$
Sm	0.5×10^{-3}	$Sm(OH)_3$
Eu	0.1×10^{-3}	$Eu(OH)_3$
Gd	0.7×10^{-3}	$Gd(OH)_3$
Tb	1.4×10^{-3}	$Tb(OH)_3$
Dy	0.9×10^{-3}	$Dy(OH)_3$
Ho	0.2×10^{-3}	$Ho(OH)_3$
Er	0.9×10^{-3}	$Er(OH)_3$
Tm	0.2×10^{-3}	$Tm(OH)_3$
Yb	0.8×10^{-3}	$Yb(OH)_3$
Lu	0.1×10^{-3}	$Lu(OH)_3$
Hf	$< 8 \times 10^{-3}$	–
Ta	$< 3 \times 10^{-3}$	–
W	0.1	WO_4^{-2}
Re	8×10^{-3}	–
Au	1×10^{-2}	$AuCl_2^-$
Hg	0.2	$HgCl_4^{-2}$, $HgCl_2$
Tl	1×10^{-2}	Tl^+
Pb	3×10^{-2}	$PbCl_3^-$, $PbCl^+$, Pb^{+2}
Bi	2×10^{-2}	–
Rn	6×10^{-13}	Rn (gas)
Ra	1×10^{-7}	Ra^{+2}
Th	$< 5 \times 10^{-4}$	$Th(OH)_4$
Pa	2.0×10^{-6}	–
U	3	$UO_2(CO_3)_3^{-4}$

- A dynamic mixing system, in which composition changes take place partly from internal processes and partly as a result of the circulation and mixing of water masses
- A reservoir that is intermediate between the runoff of components from the continents and exchange reactions with the sediments
- A biological system in which virtually all the biochemical changes associated with living organisms take place
- A grand septic tank in which organic materials are decomposed, mostly in the near-surface water, or at the bottom

- A vast chemical system, in which interactions occur among an enormous number of components, both organic and inorganic, ranging in concentration through 12 orders of magnitude
- An environment that is being invaded by man through his social, agricultural, and industrial activities
- The key to interpreting the past history of the earth and as the custodian of its relics

Very few of the properties and processes of the ocean do not involve its chemical composition. Even its movements, which may indeed not be accompanied by a change in composition, are generally observed and measured by the physical oceanographer by the distribution of its mass, calculated from salinity and temperature data. To observe movements and mixing more directly, geochemists have proposed the use of tracers such as radionuclides (for instance, ^{90}Sr, ^{137}Cs, ^{14}C, tritium, ^{32}Si, ^{226}Ra, Rn), dissolved gases (e.g., O_2, N_2, Ne, A, He), total-C, and many others, including trace elements.

It was said by the late Fritz Koczy of the University of Miami:

> Chemical reactions in the ocean . . . are largely determined by phenomena which occur at interfaces . . . seawater is bounded by two of the most extensive interfaces on earth—the one where it meets the air above, the other where it mingles with the sediments below.

The upper and lower surfaces of the ocean—the naviface and the epichthon—are as important to its material economy as the water itself. Even in the water body, the principal chemical reactions take place at the surface of either living organisms or detrital or inorganic solids such as clays and authigenic materials.

Some material is contributed to the ocean from the atmosphere and from outer space, while other substances escape from the ocean into the atmosphere. These exchanges often involve chemical alteration and isotopic fractionation. The ocean would be vastly different, and life in it impossible, in the absence of the atmosphere—were that physically conceivable.

The relatively stable composition of seawater is maintained, in part, through exchanges with the sediments of the ocean floor. Rivers, for example, contribute a major part of the dissolved and particulate matter in the oceans. However, the constituents of the riverborne material are not present in the same proportions as they are in seawater. Moreover, on mixing with seawater, some of the particulate matter in the river water may dissolve, and some of the dissolved matter may be precipitated. Thus, were it not for stabilizing factors—such as exchanges with the ocean sediments—the composition of seawater would not long maintain its present uniformity.

The most important chemical changes taking place in the body of the ocean are those associated with life processes: the photosynthetic production of organic matter; excretions and other results of organic metabolism; death and breakdown of organic matter into simpler forms, both organic and inorganic; and the recycling of much of the product of this breakdown into new living forms. These processes, and many others biochemically associated with them, affect the chemical composition of water itself, while the water's composition, on the other hand, is a determining factor in its ability to support life in the first place.

One of the oldest problems in oceanography concerns the contrast between the fertility of some areas and the barrenness of others. This problem has been largely solved, but questions still exist. We know that one requisite for fertility is an adequate concentration of common inorganic nutrients, or fertilizing elements—nitrogen, phosphorus, and silicon—but extremely small amounts of trace elements and organic substances are also necessary. There are physical mechanisms that can bring this material up from the reservoir of the deep sea, where it is abundant. Where such upwelling occurs, the sea is fertile and productive.

Most of these matters are discussed in greater detail in the chapters dealing with processes at the interfaces between the water and the atmosphere, the lithosphere, and the biosphere. Some special attention is given to analytical and other chemical methods.

While it is our purpose to give the reader some notion of the present state of knowledge in the field, this report is not a scientific review in the customary sense. We have concentrated on the knowledge itself rather than upon how it was derived, or who was responsible for it. We have not felt responsible for documentation, and, consequently, there is only a brief bibliography. Our goal is chiefly informative and introductory. The reader who would go beyond may use the bibliography as a guide.

2

Historical Outline

It is not surprising that early chemists became interested in the saltiness of seawater. Boyle, Lavoisier, Gay-Lussac, and others wrote of their experiments to isolate and assay the salts from seawater and occasionally speculated on their possible sources. Systematic chemical exploration of the ocean ("chemical oceanography," to the later professionals) began with Marcet in the 1820's but was not carried out adequately until the *Challenger* expedition of 1872–1876. Samples of seawater from all over the world were brought back and were analyzed more completely for their principal constituents than was again attempted until the 1960's. This work, carried out by Dittmar, fully confirmed the principle of constant composition, sometimes called Dittmar's law (or Marcet's principle), which has served as a basis for work in chemical and physical oceanography up to the present. According to this principle, the proportions of the major constituents of seawater are constant within a very small range of variation. Thus, it is possible to determine the total concentration of the constituents of seawater—"salinity"—by an analysis of any one of them. In practice, the constituent analyzed is generally chloride. The conception of salinity, determined in this way, has served the needs of physical and chemical oceanography very well. Recently, improved methods have brought about some modifications and have carried us into a range of precision where the exceptions to Dittmar's law are of great interest in interpreting marine phenomena.

By 1900, further progress in physical oceanography required an accurate quantitative method for determining the total concentration of salts in seawater. The International Council for the Exploration of the Sea, responding to this need, developed an empirical, operational definition of salinity from a similar empirical, operational concept of chlorinity (as described on page 13).

At that time, chemical oceanography, if there was such a thing, consisted of the routine analysis of seawater for chloride by means of the Mohr titration, an operation that could scarcely hold the attention of an imaginative chemist. With time, this analysis was assigned to technicians who could be trained to carry it out rapidly and with great precision.

Subsequently, oceanography became principally a biological science with applications in the field of commercial fisheries. Most interest in the ocean centered upon the creatures that live there and on the factors (both chemical and physical) that control their growth.

In the 1920's, methods began to appear for the determination of nutrient substances—compounds of nitrogen and phosphorus. The development took place principally in the biological laboratory in Plymouth, England, with work by Atkins and Harvey, followed by Cooper. Methods for determining phosphate and nitrate (later nitrate and ammonia), developed principally at the Plymouth laboratory, soon spread throughout Europe and to America, aided by the efforts of the International Council for the Exploration of the Sea. From then until the early 1940's, chemical oceanography was principally concerned with biological problems, as illustrated in Harvey's book, *The Chemistry and Fertility of Seawater* (1955). It was established that the conditions for plant growth in the sea are much like those on land, with both fertile and unproductive areas. Trained chemists were few in the field of chemical oceanography, although seawater analyses were being performed by other workers who would have described themselves as biologists. Work along this line progressed principally in England, Germany, the Soviet Union, Japan, France, and Scandinavia. There was some early work in Spain and Portugal as well.

During the 1920's, oceanography also began to develop in the United States. Chemical work was begun by Moberg at the Scripps Institution of Oceanography and by Thompson at the University of Washington. The establishment of the Oceanographic Laboratory of the University of Washington in 1931 expanded Thompson's activities, and for some years this laboratory was the most productive center of chemical oceanography in the United States. Thompson and his colleagues were principally engaged in the development of analytical methods, both those that could be carried out at sea and precise laboratory methods used in establishing the exact ratios of other ions to chloride. At the Woods Hole Oceanographic Institution (established in 1930) chemical work began with the efforts of Redfield, Seiwell, and Rakestraw.

This was the formative and developmental stage of chemical oceanography. A prominent figure in this development was Wattenberg in Germany. His chemical work on the *Meteor* expedition from 1925 to 1927 was the foundation for much future development. The data he accumulated from the South Atlantic for the distribution of pH and such constituents as phosphate, ammonia, calcium, and carbonate were widely used for many years.

During the 1930's, analytical methods began to appear for other constituents of seawater, such as iron and copper, and studies were begun on the possible relation of such trace metals to the growth of organisms. Other elements, found in very low concentrations, were soon added to the list of constituents. A beginning was made in the study of dissolved organic materials in seawater, the importance of which was later to become more fully appreciated.

Although until 1940 the principal use of chemical oceanography was in the attack on biological problems, there were a few other areas of interest to chemistry. One of these was the exchange of carbon dioxide between the sea and the atmosphere. Buch in Finland was the principal contributor to this subject, assisted by others. Although the physical chemistry of carbon dioxide and of solutions of it in pure water was fairly well known, data for its solutions in seawater remained to be accumulated. An important experimental study by Buch, aided by Harvey, Wattenberg, and Gripenberg, determined the necessary constants and became the sole source of such information until very recent times.

By 1940, the complexion of chemical oceanography had changed notably. Marine geology, or the geological aspects of oceanography, had been developing through the previous decades, and it had become quite evident that the chemistry of seawater was fundamentally involved in sedimentation phenomena. Understanding the formation of calcium carbonate is a key probelm in studies of the marine sedimentary cycle, and the chemical environment in the ocean in which it takes place had been under investigation for some time. The work of Wattenberg on the *Meteor* expedition suggested that some areas of the sea were supersaturated with calcium carbonate, while other areas were undersaturated. This study of the calcium carbonate problem was one example of interest in the chemistry of the sea outside the biological area.

Another major division appeared—marine geochemistry—concerned not merely with the use of chemistry to solve geological problems, but also with the part the ocean plays in the broad, general weathering cycle. It therefore dealt with the distribution of any and all of the chemical elements in the sea. Since most of the chemical elements have been found in seawater, the chemist is provided with an endless number of problems concerning the source, speciation, function, and significance of these elements and their interactions.

Another development of considerable importance to chemical oceanography was the advent of atomic energy, which led to the development of many useful scientific techniques. The utilization of radionuclides in the marine field has enormously affected chemical oceanography, the new methods of study opening whole new fields of investigation. Radioactive isotopes, both natural and artificial, are being widely used in oceanographic studies as tracers and time clocks. Since a very detailed report dealing with this subject, *Radioactivity in the Marine Environment*, has recently been published by the National Academy of Sciences, this report will present only a limited discussion of the field.

With the development of this field, studies were also begun on the stable isotopes, such as those of hydrogen, oxygen, carbon, and sulfur. These isotopic species have been effectively used in physical and biological investigations and have given us some clues to the origin of the ocean itself.

The last two decades have witnessed an enormous increase in marine chemical research activity, partly as the result of the entry of many new workers into the field and partly as the result of the appearance of new techniques and methods of analysis. The flexibility and precision of the new approaches have made possible an attack on problems that, in earlier days, could not have been attempted, the results of which have opened up entirely new fields.

A substantial beginning has been made, for example, on the more intimate investigation of the composition of seawater and of the speciation or molecular structure of its many components and of the interactions between them.

3

Physical Chemistry of Seawater

COMPOSITION AND STRUCTURE

Composition of Seawater

During the past few years there have been comprehensive studies of the chlorinity ratios of the major constituents in seawater. These investigations have refined our knowledge of the composition of seawater and have set upper limits to the variability of most of these constituents. It appears that any variability in the ratios of sodium, potassium, magnesium, sulfate, or bromide to chloride has been known with analytical precision for some time. There is conflicting evidence regarding the variability of strontium in the ocean. Variations in calcium with depth have been well defined. High fluoride–chlorinity ratios have been observed in the deep North Atlantic. There have been conflicting views regarding the occurrence and significance of organically bound boron in the oceans, and further study is needed.

Progress has been made in the determination of the major chemical species in seawater. Values for the extent to which sulfate, carbonate, and bicarbonate form ion pairs have been reported, and further measurements are being conducted. Experimental studies are needed regarding the speciation of elements in moderate concentration, such as potassium, boron, strontium, and fluorine.

Except in a few instances, the analytical problems invovled in the determination of the minor elements are so severe, and so little work has been done, that we are uncertain of the vertical or horizontal variations of these elements. We need more comprehensive studies of the behavior of the minor elements, including their interactions with organic and inorganic matter and other relations that may influence their concentrations.

Chlorinity–Salinity Relation

Chemists require a value for the concentration of seawater when studying the thermodynamics and kinetics of chemical reactions and of mass and energy exchange. However, it is virtually impossible to measure directly the total dissolved content, for any effort to recover and weigh it is accompanied by decomposition and loss. It is possible to obtain a constant weight only under arbitrary operational conditions.

As pointed out in Chapter 2, because of the apparently constant relative composition of seawater, it appeared possible to determine the total dissolved content by careful measurement of the concentration of any one of the principal ions—most appropriately chloride. The relation of this earlier parameter, "chlorinity," to salinity was established by accurate measurements on a very few samples of shallow seawater and, in its original form, $S = 1.805\text{Cl} + 0.03$, neglected possible regional and depth variations in the relative composition.

A new relation, $S = 1.80655\text{Cl}$, was recently proposed. This equation simplified the use of conductivity measurements in determining salinity and preserves closely the original S–Cl relation.

Measurements of conductivity have largely replaced the older chloride titrations, being generally more convenient and more precise. Salinometers for this purpose, including designs for *in situ* chlorinity determination, have been developed.

Conductivity would seem to be more directly related to total concentration than chlorinity is, and more responsive to changes in relative composition. Nevertheless, in order to retain continuity of data, salinity is still referred to chlorinity, and use is made of conductivity–chlorinity ratios. The new relation of chlorinity to salinity, like the old, does not recognize the possibility of variations in the relative composition of seawater.

Such a "salinity" is merely a defined quantity. However, it can be derived from its relation to chlorinity or conductivity and may very well be used as a state variable if it is shown that its relation to conductivity or chlorinity provides a good enough first approximation to the ionic composition of seawater. There will continue to be much thought given to the definition of salinity and its usefulness as an oceanographic parameter.

After all, salinity is not strictly necessary in the study of water masses and for geostrophic computation. For these purposes, the specific gravity or specific volume is the more desirable parameter. Uncertainty as to the precision of the relation of specific gravity to chlorinity has led to the requirement of direct measurement of the specific-gravity–conductivity relation, and work on this is now under way.

The desired ultimate relationship, of course, is a direct conversion of conductivity to absolute density. Its establishment involves many difficulties. For example, a knowledge of the isotopic composition of the water used for calibration is an essential prerequisite. Progress is being made in this direction.

Water–Ion Interactions

The properties of seawater differ from those of pure water because of the "dilution" of the water by dissolved constituents, and because of the strong interactions between water molecules and the ions of the dissolved salts, as well as because of ion–ion interactions. Much of our knowledge of ion hydration in salt solutions has been derived from studies of conductance, vapor pressure, and freezing-point depression. If the structure of pure water were well known, the way would be open to interpretation of salt effects on a molecular scale. It appears likely that a major effort in chemical oceanography in the future will be the unraveling of the fine structure of seawater, moving hand in hand with general advances in the theory of concentrated salt solutions. At present our knowledge is primitive, and the nature of water–ion interactions, particularly in the case of ion pairs, is just beginning to be defined. Knowledge of hydration numbers and energies, structures, and rates of reaction would go far toward providing a background for a comprehensive theoretical framework of the physical chemistry of seawater.

THERMODYNAMICS

Bulk thermodynamic properties of seawater, such as specific volumes and specific heats, are useful for the physical description of oceanic processes, such as adiabatic mixing. The chemist, however, is primarily interested in the contribution to the thermodynamic properties of seawater of its individual constituents, organic and inorganic. Until now, only the partial molal volumes have received thorough attention; little work has been done on heats of mixing and free energies of mixing. In view of the importance of the vapor pressure of seawater, the differences between the data of the various investigators need to be reconciled.

The pressure and temperature dependence of chemical equilibria deserves greater attention. At present, such equilibria can be described only by means of apparent constants or by means of thermodynamic constants with apparent activity coefficients. The chemistry of solutions of high ionic strength and of the complicated nature of seawater has scarcely been touched. Systematic studies in mixed electrolytes, less complicated than seawater, and in seawater itself, should eventually yield the necessary information for a rigorous thermodynamic description of chemical equilibria in seawater. For questions as complicated as this, the theoretical foundation currently available to solution chemists may not be sufficient, and formulation of an adequate theoretical foundation presents a great challenge. A knowledge of such properties as excess free energies, heats of mixing, and excess free entropies will be imperative.

Thermodynamic studies of the effect of the dissolved salts reactivity of nonelectrolytes in seawater are called for. The effect of the ionic strength upon the rates of simple inorganic reactions in seawater has also been neglected. Nonequilibrium thermodynamic studies (of thermoelectric diffusion, for example) will be of great importance, especially in the investigation of interstitial waters.

In principle, equilibrium constants can be used to determine the chemical species present in seawater for reactions that are fast relative to hydrographic, biological, and geological processes. The speciation must be known, because it affects the charge and particle size distribution and, therefore, the physicochemical properties of seawater. For slow reactions, departures from equilibrium can be used, and factors that prevent equilibration may be sought. Examples of slow reactions that need to be studied further are clay alterations and interactions with silica and with metal ions, the solution of carbonates at depth, the exchange of gases at the sea surface, and the oxidation of organic matter.

Two types of equilibria need to be considered:

Dissociation of Weak Acids The dissociation of carbonic, boric, and phosphoric acids at high pressures has been studied in seawater. A limited amount of work has been done on their temperature dependence. The next step should be obtaining high-pressure data on such important species as phosphoric acid. Atmospheric pressure measurements of the dissociation of hydrogen sulfide, ammonia, amino acids, organic acids, silicic acid, and clays would also be useful.

Ionic Association and Complexing Sodium, calcium, magnesium, and potassium are known to form ion pairs with sulfate, carbonate, and bicarbonate in seawater at 25°C and atmospheric pressure. The extent of such association at other temperatures and pressures found in the marine environment will eventually be measured.

The association of chloride and hydroxide with several trace metals in seawater has been postulated. Study of this association is a first step in understanding the significance of such complexes in the ocean.

In general, information on the binding of ions present in the oceans to each other, to clays, and to organic matter is scarce. This information is important for the study of biochemical activity and of the fate of pollutants, for understanding the physicochemical properties of seawater, and for the recovery of materials from the sea.

RATES OF REACTIONS

Little has been done in the study of reaction rates in seawater. Such work is needed to explain the presence and to predict the fate of chemical assemblages that are not at equilibrium.

Some important species that have been reported to occur in the ocean, even though they are thermodynamically unstable, are manganous, iodide, and arsenate ions. Possibly their presence is due to the lack of reaction sites needed to bring about equilibration.

More work is needed on the rate of solution of carbon dioxide under various conditions and on the rates of clay–silica and clay–metal interactions.

PHYSICOCHEMICAL PROPERTIES OF SEAWATER

Colligative Properties

The boiling-point elevation, melting-point depression, vapor pressure, and osmotic pressure of seawater have been the focus of several recent studies, mainly aimed at refining the results of previous investigations by more sophisticated experimentation and instrumentation. A knowledge of these parameters allows more extensive thermodynamic treatment of seawater systems and has applications to the industrial utilization of seawater, heat-exchange processes, and desalination procedures.

Viscosity

Studies of the viscosity of seawater have been concerned especially with its temperature and pressure dependence. In the vertical water column, viscosity varies as a result of temperature and pressure differences and biological activity. Efforts are being made to extend the precision and range of measurements of viscosity as a function of temperature and pressure.

To the chemical oceanographer, this is also a useful parameter in studying water structure, hydration of ions, and ionic complexes. Therefore, increased precision in measurements of viscosity can lead to important new knowledge in these areas.

Specific Heat

The specific heat is needed for the calculation of the adiabatic temperature gradient, heat budgets, and the heat of mixing of waters. It may also be a useful parameter in studies of the molecular structure of seawater.

Specific Volume

This property, of great importance to the physical oceanographer, has already been referred to, in its relation to salinity, chlorinity, and conductivity. This is one of several properties of seawater that, although primarily of physical interest, are of concern to the chemist because of their dependence on the chemical composition of the water and their consequent involvement in the uncertainty of the salinity–chlorinity–conductivity relation.

Electrical Conductivity

We have already dealt with the relation of conductivity to chlorinity, which it is rapidly replacing as an operational parameter. "Conductivity ratios" are determined by relative rather than absolute measurements.

Electrical conductivity is also a tool for the study of the structure of water and the speciation of ions in solution. Most of the work in this field has been done with simple solutions and now needs to be extended to the more complex system of seawater itself.

Partial equivalent conductances, already obtained, will help in dealing with the complexities of the conductivity–specific-volume relation.

Sound Absorption

Sound absorption in seawater, important in detecting underwater objects, is greater than in pure water at frequencies between 10^4 and 10^6 Hz. This effect has been shown to be due to the presence of magnesium sulfate in seawater. The degree of association between magnesium and sulfate ions in seawater has been calculated from sound absorption measurements, and a good agreement was obtained with the results of other techniques.

Sound absorption has also been found to be anomalously high in seawater and in fresh water at frequencies between 10^1 and 10^3 Hz. It has been suggested that the large absorption at low frequencies might be due to a structural relaxation of water. However, the effect of chemical relaxations, particularly in the carbon dioxide system, may also be responsible.

4

Air-Sea Interaction

Three states of matter are found at the air–sea interface: solids, such as dust; liquids, such as rain; and gases, such as the atmospheric gases. Whereas the exchange of energy and of water between ocean and atmosphere is principally responsible for the influence of the ocean upon the world's climate and weather (and therefore of primary interest to the meteorologist), there are other exchanges involved in many oceanic phenomena that are of primary interest to the chemist. Of these, the exchange of atmospheric gases is the most evident.

ATMOSPHERIC GASES

All of the atmospheric gases exist in the sea. While equilibrium conditions are approached, though perhaps only occasionally attained, it is nevertheless usually assumed that the concentrations of dissolved gases are governed by solubility coefficients that are functions of temperature, salinity, and pressure. There are indications, however, that kinetic or other restraints may actually prevent the attainment of equilibrium between the sea and the atmosphere at the sea surface.

Oxygen

Except for chloride, there is no other substance whose concentration and behavior in the ocean has been more widely studied than dissolved gaseous oxygen. The reason is twofold: It is relatively easy to determine the concentration of oxygen analytically; and this quantity may be more indicative of the properties and previous history of a seawater sample than any other. Because of the involvement of oxygen in many of the biological and inorganic processes in the sea, variations in its concentration may be diagnostic of much that has gone on, as well as indicative of what may be chemically possible.

The solubility coefficients as functions of temperature and salinity have recently been determined to a high degree of accuracy by three laboratories (MIT, Johns Hopkins, and the University of Liverpool). On shipboard, the Winkler method of determination has been used for many years, with results entirely satisfactory for solution of the problems of yesterday. However, minor oversights, now recognized, have limited its accuracy. Present and future demands for greater accuracy may necessitate continued improvement of this method and perhaps the development of others.

The oxygen found in seawater has two sources: exchange with the atmosphere and production in the euphotic zone through photosynthesis. The isotopic compositions of the oxygen from these two sources are different and can, in principle, be used to measure their relative contributions. Although some initial work on this was done more than a decade ago, it has not been vigorously pursued.

The distribution of oxygen in the ocean is characterized by certain general features:

1. Near-saturation at the surface—deviations from saturation may result from
   a. Rapid changes in temperature
   b. Upwelling from either the photic zone, resulting in supersaturation, or deeper water, producing undersaturation
2. Decrease in concentration through the thermocline to a minimum
3. Increase below this minimum to a value that may be greater than that at the surface immediately above

Secondary maxima and minima are not uncommon in the vertical profile. These features are shown in Figure 1, which includes two extreme cases: (1) in which oxygen is nearly constant in the vertical

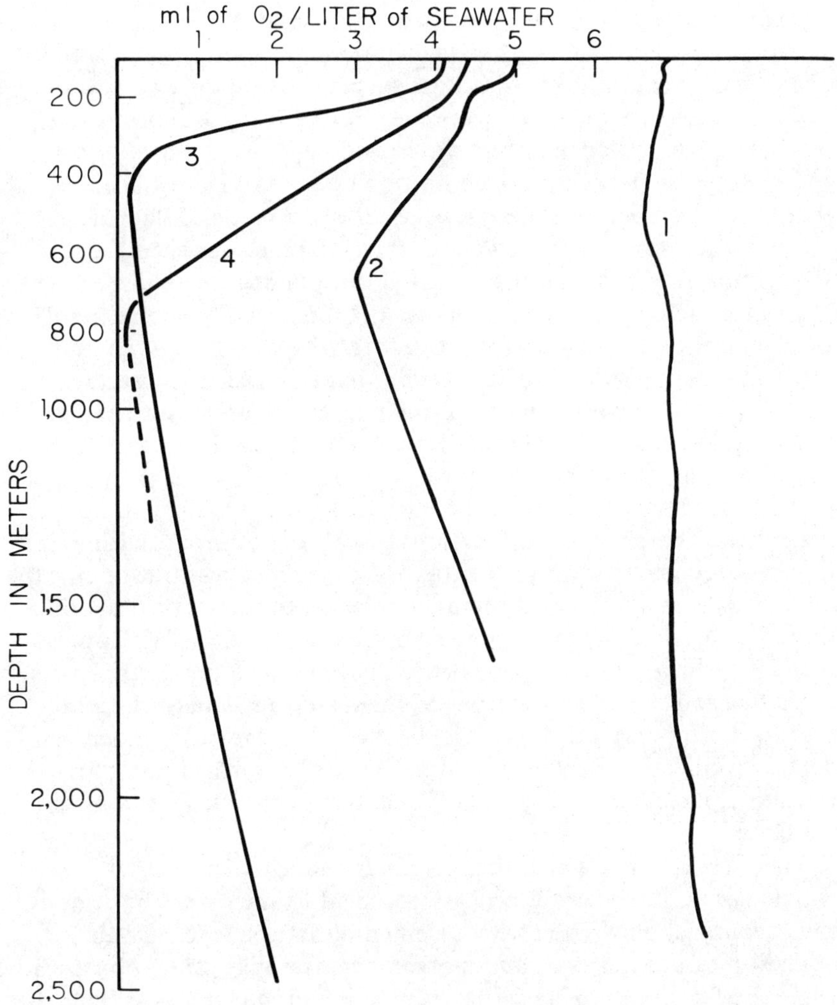

FIGURE 1

column (presumably from the local sinking of surface water), and (4) in which oxygen is completely lacking at a mid-depth below which H_2S appears from the reduction of dissolved sulfate. The observed pattern of oxygen distribution is principally dependent on the occurrence and extent of photosynthesis, respiration, and the oxidation of organic matter, superimposed on the movement of water, both horizontal and vertical. The distribution of organisms is a highly important contributing factor; temperature, probably less so.

The commonly observed "oxygen minimum layer" has long been a matter of great interest. Although we can understand and appreciate the chemical and physical processes that bring it about, we do not have an adequate quantitative theory for its existence.

In interpreting oxygen concentration as an indicator of changes that have taken place in the water, use is often made of "apparent oxygen utilization," AOU, which is the difference between the observed concentration and that which corresponds to equilibrium with the atmosphere at the observed potential temperature and salinity, on the assumption that the water was saturated somewhere at the surface in the past. While it seems certain that this difference results chiefly from the oxidation of organic matter, we are unsure of the nature of this material—whether it was living or dead—or, indeed, just where the oxidation took place.

An early theory postulated a continuous rain of living and dead organic matter from the surface to the bottom of the sea, with oxidation proceeding meanwhile. Neither the oxygen concentration nor the AOU, however, show any systematic relation to the vertical distribution of either dead or living organic matter, and it therefore appears that the utilization of oxygen is virtually limited to the upper layers of the sea and that other features of the distribution pattern result from water movements. Nevertheless, there is some independent evidence for oxygen consumption in deep water, a vertical rain of particulate organic matter does take place, but the whole picture is not yet clear.

There are areas in which the oxygen is entirely depleted—the anoxic zones. Little attention has been paid to such areas by American oceanographers. Even where elemental oxygen is totally absent, oxidation of both organic and inorganic matter may still go on at the expense of such oxidizing agents as nitrate, sulfate, and possibly carbonate. This gives rise to such reduction products as nitrite, elemental nitrogen, hydrogen sulfide, and possibly ammonia and methane. Nitrite also occurs in the presence of oxygen and may result from processes other than the reduction of nitrate.

Nitrogen

There is little evidence for any considerable fixation or production of nitrogen in biological processes in the ocean; nitrogen concentrations in seawater are very close to saturation values. Molecular nitrogen is a product of oxidation processes in anoxic waters, but it appears that neither this process nor that of nitrogen fixation is of major importance.

Noble Gases

The noble gases in seawater are of interest even though they are not involved in biological activity. Indeed, information about them is of use for that very reason.

If it may be assumed that the water was originally saturated with the noble atmospheric gases at the surface at some time in the past, and that they remain unaffected chemically, then the observed concentration of these gases in deep water may give a clue to the history of the water since the time of saturation. If the observed concentration of the gas is compared with its solubility at the temperature and salinity of the water *in situ*, the difference may be either positive or negative. We may then ask what physical changes within the water can have caused this difference, as for example, mixing of water masses, changes in temperature, transport of air bubbles from the surface, and, of course, the advent of new gas from other (radioactive?) sources. The solubility curves of gases are generally concave, and this results in supersaturation when saturated water masses are mixed. This effect, however, varies for different gases.

Therefore, by determining the ratios between the concentrations of noble gases, and their comparative deviations from saturation, one may be able to decide which of the several possible physical changes has taken place in the water. Present work indicates that measurable deviations from saturation do occur, resulting in both supersaturated and undersaturated waters. Although some of these results can be related to mixing processes, others appear inexplicable in the absence of more accurate solubility data.

Helium is a special case, because of its possible input into deep water from radioactivity in the solid earth. Indeed, excess helium levels have been reported in both the Atlantic and Pacific oceans by a number of laboratories.

Isotopes of radon, introduced at the water–sediment surface by the decay of radium in the lithosphere or produced in the water through such decays, are among the tracers for water masses and mixing processes.

Carbon Dioxide

The interchange of CO_2 at the sea surface is more complicated than that of most other gases because it is accompanied by chemical combination, with the formation of bicarbonate. The "carbon dioxide system" CO_2-HCO_3^--CO_3^{-2} consists of both gaseous and dissolved components and is important both biologically and geologically, bearing on photosynthesis on the one hand and carbonate sedimentation on the other. The physical chemistry of this system is well known, but there are factors in its application to the marine environment that are still obscure. Carbon dioxide passes from the atmosphere to the ocean more slowly than does oxygen.

The possibility that the burning of fossil fuel may be increasing the concentration of carbon dioxide in the atmosphere, and that this may eventually influence climate, has led to many measurements of atmospheric CO_2 tension by infrared absorption. For determining the partial pressure of CO_2 in seawater, this method is much more sensitive and accurate than calculations based on chemical determinations of pH and titration alkalinity and the use of empirical dissociation constants. The newer techniques have made it possible to show that the carbon dioxide content of the atmosphere is increasing by about 0.7 of a part per million per year—about half the rate at which it is introduced into the atmosphere by the burning of fossil fuel. The remainder is evidently being taken up by the ocean as well as by the rest of the biosphere. The geographical and other factors involved are still under study.

The quantitative balance of carbon in the ocean may turn out to be as important for the oceanographer to consider as that of oxygen, which has long occupied his attention.

Kinetics

The mechanisms for gas exchange reactions (invasion and evasion coefficients) are poorly known. Rates of exchange vary somewhat for the different atmospheric gases and are very much slower for very soluble gases, such as CO_2. Evidence seems to indicate that the controlling factor is the rate of diffusion within the solution, although processes and conditions at the immediate surface of the ocean (collision efficiency, for example) may also be responsible. In the special case of CO_2, intermediate enzyme action has even been postulated.

OTHER GASES

Carbon Monoxide

Some recent work indicates that the oceans may be a source of atmospheric carbon monoxide, perhaps through the incomplete oxidation of organic matter in seawater or as a metabolic product of organisms. Until these initial measurements were made, it was thought that the oceans might be a sink for the large amounts of man-made carbon monoxide. Much work is needed in this area.

Hydrocarbons

Among the gaseous trace substances in seawater are the low-molecular-weight hydrocarbons. Their sources and significance have not yet been ascertained. Some are clearly related to the activities of man; others may result from natural processes. But they may provide most useful diagnostic tags for biological processes.

Hydrogen Sulfide System

Hydrogen sulfide is formed from the reduction of sulfate in waters depleted of gaseous oxygen. The simultaneous occurrence of both oxygen and hydrogen sulfide in a given water mass, although reported, is not as yet established. The system has been studied in such areas as the Black and Baltic seas, the Cariaco Trench, Walvis Bay, and Scandinavian and British Columbian fjords.

Alterations in the composition of seawater resulting from the precipitation of heavy metals and the accumulation of strong metal sulfide complexes in anoxic zones have been investigated. There is a marked fractionation of the sulfur isotopes in favor of ^{32}S over ^{34}S in the formation of pyrite and other iron sulfides from sulfate reduction. A study of such isotopic changes may be useful in defining the characteristics of ancient marine environments.

Ammonia

The occurrence of ammonia in bodies of water comes about principally from protein decomposition, but there appears also to be an atmospheric input via rain of ammonia resulting from similar processes in the metabolism of land animals. Ammonia occurs only in near-surface or surface waters and is rapidly oxidized to nitrate or utilized directly by phytoplankton.

Radioactive Gases

Three cosmic-ray-produced radioisotopes, ^3H, ^{14}C, and ^{39}Ar, enter the oceans from the atmosphere. They are returned to the atmosphere from the oceans as the gaseous molecules $^3H^1HO$, $^{14}CO_2$, and ^{39}Ar. In principle, a knowledge of their atmospheric specific activities can be used to introduce the time parameter to mixed-layer waters prior to their sinking to depths below the thermocline. Some preliminary results with ^{14}C in eastern Pacific water masses have allowed calculations of their rate of movement.

The use of the first two of these isotopes has been complicated by their production in the detonation of nuclear devices. By this means, the tritium concentration in the atmosphere has been increased by an order of magnitude during the last decade or two. This effect has allowed studies to be made of the residence times of water molecules in the atmosphere, troposphere, and surface mixed layers of the ocean.

In addition, the rates of carbon dioxide invasion and evasion through the ocean–atmosphere interface have been obtained through measurements of carbon-14 in the atmosphere and in the surface waters. The usefulness of such radioactive materials in introducing the time parameter to marine systems is just now being exploited.

SOLIDS

Dusts

Continental and volcanic solids are introduced into the marine environment through atmospheric transport. Such major wind systems as the jets and the trades, at middle and low latitudes, respectively, provide a significant portion of marine sediments through the fallout of their dust burden. Extraterrestrial materials, the cosmic spherules, are found in the sediments in extremely small amounts. Man-introduced solid materials are appearing in the air, water, and sediments and may provide important tracers in studies of marine sedimentary cycles.

Aerosols

Dissolved species in seawater are thrown into the atmosphere by evaporation of sea-spray during turbulent activity in surface waters, a process often called "splitting" of seawater. Such materials are returned to either the land or ocean surfaces primarily by rain or snow, although dry fallout accounts for about 25 percent of the return. Nearly all of the chlorine and about half of the sodium in river waters result from such a transfer.

Very little is known about any elemental fractionations that may take place during this splitting of seawater, although there are often conflicting accounts in the literature describing the enrichment or depletion of one ion relative to another. The markedly high concentrations of iodine, fluorine, and boron in rains, compared to seawater (when normalized to chlorine), may be due to a splitting phenomenon or to a volatilization, or both.

5

Marine Biosphere

In the study of the marine environment, as on land, the traditional disciplines of chemistry and biology are not separate, but continuous. Nonetheless, in order to view the status of marine chemistry in relation to biologically organized matter, both living and dead, some dividing line must be drawn as a matter of practical necessity. A useful approach is to consider the biological chemistry of the sea as an integral part of geochemistry, and this approach has been taken here. We have paid particular attention to events that take place over relatively short intervals of geologic time and that are largely under the control of biological agencies.

CHEMICAL CONSTITUENTS OF THE MARINE BIOSPHERE

A great deal of information has been gathered on the protein, lipid, and carbohydrate composition of whole marine organisms, particularly plankton. The situation with regard to more detailed analysis of their constituents, such as carbohydrate sugars, amino acids, and fatty acids, is more irregular. Here, comprehensive inventories are not available, and studies of the peculiarities of chemical composition of various organisms and their relation to variations in the composition of the water in which they live seem to be rare. There appears to be considerable potential for development in this area. As far as specific substances or families of substances are concerned, only limited collation of data has taken place in recent years, although analyses of a wide variety of compounds have been made by individual laboratories. In regard to the algae alone, for example, it is possible to find data dealing with carbohydrate and protein constituents, free amino acids and peptides, plankton fatty acids, sterols, bromphenols, DNA, gibberellins, antibiotics and toxins, and many others. However, little information is available on the composition of whole tissues or cell organelles—information that might well be required in particular circumstances.

There has been a long-standing interest in the chemical composition of nonliving, particulate, and dissolved organic materials in the sea. Where small organisms are present, it is impossible by means now available to separate the living from the dead particulate material. This problem persists in the face of serious efforts in a number of laboratories to seek a reliable basis of distinction, such as using carbon–chlorophyll ratios.

Analyses for total dissolved and particulate organic carbon are now available for most of the major water masses. Some controversy has arisen over the interpretation of such data, however. It has been argued that the organic matter in the deep sea is in a continuous state of flux; on the other hand, there is evidence that this material is largely nondegradable or poorly degradable, representing an accumulation of compounds over an extended period of time. Important questions yet to be answered relate to the modes and location of production of organic compounds entering the deep sea, their fate, and their influence on associated chemical variables. The greatest drawback to an easy solution of these problems has been the lack of suitable techniques for analyzing the variety of organic materials involved, including both solid and dissolved forms. To date, only some 10 percent of the organic compounds present in seawater have been identified. With the establishment of more sensitive and selective analytical techniques, data, not only on carbon, but on specific compounds are beginning to accumulate.

Chemical differentiation of marine organic particulates has received less than adequate attention. A study of carbohydrates in suspended material taken in profiles off Japan indicated that 1,3-glucan and related polymers disappear at depths greater than 50 to 300 m, leaving other water-insoluble carbohydrates that are resistant to biological attack. More extensive data of this kind are urgently needed to resolve difficulties in the interpretation of the depth distribution of dissolved and particulate carbon. It has been reported that marine organic particulates are often much richer in nitrogen than regular humic materials. An extensive Pacific survey revealed several physically distinct classes of organic detritus. Unfortunately, chemical analyses were not attempted.

Qualitative tabulation of the known dissolved organic compounds is extensive and includes carbohydrates, proteins and their derivatives, aliphatic carboxylic and hydroxy-carboxylic acids, humic acids, phenolic compounds, and hydrocarbons. Although this list may appear impressive, many compounds remain to be identified—most importantly, those that exert controlling influences well out of proportion to their concentrations.

Quantitatively, the total amount of organic matter in the ocean is rather uniform and does not vary widely below the thermocline. On a molecular scale, however, the story may be quite different. Compounds encountered may differ greatly, depending upon their source. The body chemistry of the organisms from which dissolved organic compounds originate is highly variable but is species-specific. Organics from pollution sources, or from bacterial action, also have their own recognizable imprints. Such specific variations in molecular composition can have diagnostic value for the study of marine processes and may provide tags for water masses.

It has been pointed out that over 90 percent of cellular material includes less than fifty compounds and polymers of those compounds. Billions of possible low-molecular-weight compounds theoretically can be synthesized from carbon, hydrogen, nitrogen, oxygen, phosphorus, and sulfur, yet the *Pfizer Handbook of Microbial Metabolites* (1961) lists only 1,313 compounds, many of them restricted to a few species. Compounds regarded as ubiquitous in living material number between sixty and seventy. A study of specific deviations from the predicted natural chemistry might be of considerable geochemical interest. Yet to expect an exhaustive quantitative inventory of specific compounds would be unrealistic. Minimally, the inventory should include carbon in living and nonliving materials, with the living component subdivided by taxonomic groupings and the nonliving categorized as particulate or dissolved. It would not be unreasonable to include in this inventory lipids, proteins, carbohydrates, nucleic acids, humic materials, and substances of low molecular weight. While guiding estimates probably are available in existing literature, the information has not been collated, even to this limited extent, on a global basis.

The long-term goal of the study of dissolved organic matter in the ocean is to understand from what sources and in what manner organic compounds are introduced into the sea, what happens to them once there, how they interact with inorganic constituents of the ocean and how this knowledge can be used to gain a better appreciation of what goes on in the ocean as a whole.

The organic content of marine sediments, which includes lipids, proteins, and carbohydrates, along with often large amounts of non-hydrolyzable matter, may be considerable. Trace organics have been studied by chromatography, and there has been interest in the organic matter of ancient sediments, much of which is related to porphyrin. Although a good deal is known about the organic chemistry of the humic substances, these by no means account for the totality of organics in sediments, and a vast amount of work awaits the marine equivalent of the soil chemist.

CARBON CYCLE

We are here concerned with the material dynamics of the marine biosphere: the rates and magnitude of photosynthetic (and the relatively limited chemosynthetic) fixation of carbon from the inorganic carbonate–bicarbonate–CO_2 pool, as well as the subsequent rate of passage of this carbon through the biological system toward its ultimate return to the CO_2 pool. The problem also involves the many other elements and their compounds affected by this activity. It is further complicated by the fact that the total system maintains itself in a fluid medium characterized by constant circulation and mixing. While overall patterns are reasonably well appreciated, the documentation of detail is not very far advanced.

Various estimates of the annual rate of incorporation of carbon into marine and terrestrial ecosystems have been made over the years. The most recent indicate an annual fixation rate of 1.5 to 1.8×10^{10} metric tons, reasonably close to the values obtained in 1952 in the first wide-scale measurement of productivity in the sea using carbon-14 techniques. Our concept of the total productivity of the world ocean has apparently changed little since then. The validity of this estimate obviously is basic to any quantitative discussion of marine biosphere dynamics.

According to the figures presented in Table 2, only about 0.3 percent of the organic carbon in the oceans is associated with living organisms, and the total stock of organic carbon is minuscule compared

TABLE 2 Distribution of Carbon[a]

Compartment	Total Carbon (as CO_2) × 10^{18} g
Carbonate in sediment	67,000
Organic carbon in sediment	25,000
CO_2 in atmosphere	2.35
Living matter on land	0.3
Dead organic matter on land	2.6
CO_2 as H_2CO_3 in ocean	0.8
CO_2 as HCO_3^- in ocean	114.4
CO_2 as CO_3^{-2} in ocean	14.2
Dead organic matter in ocean	10.0
Living organic matter in ocean	0.03

[a]Reprinted with permission from Revelle and Suess (1957).

with the global reserve. Of that living stock, perhaps 90 percent exists in the material of phytoplankton. Geochemically, however, it would be unwise to neglect the apparently trivial remainder, since the rate of flow of matter through that small stock may have considerable significance. It is doubtful, for example, whether the large reserve of nonliving organics has been derived in a single step from algal cells.

CHEMICAL AND BIOCHEMICAL FACTORS

During the past 25 years, the main emphasis in the field of primary productivity has shifted from studies concerned mainly with inorganic micronutrients to studies of the organic nutrient chemistry of seawater. These investigations have led, in part, to the identification of specific growth-promoting substances and to the discovery of the presence of substrates in sufficient concentrations to support heterotrophic and chemosynthetic growth. From these observations, it appears that the properties of seawater as a medium for the support of life require extensive definition in order to explain not only the presence of autotrophic organisms but also the occurrence of large blooms of auxotrophs, heterotrophs, and other organisms that depend on an organic factor for their growth. Some advances along these lines have been made in the collection of data on vitamins (especially B_{12}), chelating compounds, specific growth-promoting substances, and the presence of autoinhibitory substances. However, most reports on organic micronutrients and antimetabolites have been written more to demonstrate the existence of a process than to show its quantitative ecological importance.

Discovery of the heterotrophic and chemosynthetic activity of certain organisms in seawater has demonstrated a need for more understanding of the level and, more importantly, the rate of turnover of organic and inorganic sources of energy in the marine environment. Potential substrates for heterotrophic growth need to be identified, and the levels of the more common substrates, such as glucose, acetate, and some amino acids, need to be determined both in relation to possible seasonal variations and in terms of the minimum concentration required to support growth. In this respect, it is now apparent that some unicellular algae exist, either heterotrophically or by chemosynthesis, at great depths in the oceans. The capacity of deep ocean water to support their growth requires investigation. It is possible that the antimicrobial activity of humic materials may be important in this regard. In addition, the report of the initially inhibitory effect of recently upwelled deep water on many photosynthetic phytoplankton requires chemical or biochemical definition. Further, the general ecological significance and the nature of the toxins and antibiotics produced by marine organisms are not well understood.

Fairly extensive geographical and seasonal data have been collected on the amount of phosphate, nitrate, and silicate in the world oceans. Routine measurement of these compounds in surface waters, over wide areas, appears to be possible through the use of continuous recorders on commercial vessels. Data on the occurrence in the oceans of other forms of phosphorus and nitrogen (such as organic N and P, ammonia, and hydroxylamine) have been less extensively collected. Future needs for the routine analysis of these compounds may develop from a better understanding of their uptake by phytoplankton, especially in oligotrophic environments. The importance of trace metals (such as Fe and Mn) in the metabolism of phytoplankton has been demonstrated. The possible seasonal variation in their availability and quantity has not been studied over large areas of ocean.

The importance of nutrients and the production of growth-promoting substances in sediments require considerable attention by both organic and inorganic chemists, and further studies are required on the importance of sediments in the release of organic metabolites. Many opportunities await investigators who are interested in the phenomena associated with the formation of organic aggregates.

The above discussion touches on only a few of the many requirements for research in this field, not including, for example, specific mention of chemotaxis, a phenomenon whose importance is suspected and, in a number of instances, documented, but for which measurements of quantitative influence are scarce.

The geochemical role of organic compounds in the sea has been receiving increased attention. In addition to organic-material interaction, organic compounds take part in various other geochemical processes. It has been reported, for example, that the lack of measurable interaction between seawater and carbonate-material particles is due to organic coating on the surface of the minerals.

A great variety of organic compounds have been found to be associated with clay minerals in ancient and recent sediments. This is a matter that particularly concerns the soil chemists and petroleum chemists.

The mechanism of the interaction between organic compounds and sediments is not well understood, and the incorporation of the organics in the sediment deposits as a function of such factors as pH, salinity, and temperature is unclear. The state of the material—whether it is physically adsorbed to the mineral surface or chemically bound to the mineral functional groups—is also poorly understood.

6

Relations between Seawater and Sea Bottom

The lower boundary of the oceans, like the upper one, is a zone of two-way transport of matter and energy. Sediments continuously rain down; overlying waters dissolve some materials from the sediments and precipitate others; pore waters are expressed upward through the compacting sediments, while dissolved species move up and out of the sediments by ionic and molecular diffusion as well as by entrainment in pore water. Diffusion downward into the sediment also can take place. Volcanic activity pours lavas out on the pre-existing sediments, releases volatiles to the local system, and creates thermal and chemical anomalies.

These processes are poorly known quantitatively, limiting our ability to develop a mass-balance of the energy and material inputs and outputs of the oceans, except in an imprecise way for a few substances.

INPUT BY STREAMS

About 85 percent of the material input to the ocean comes from streams and groundwater, and 10 percent from ice. The remainder consists of material eroded by waves and winds and cosmic dust trapped in the atmosphere. (These are estimates and their probable error is large.)

Recent data on rivers (the most important material source) show the general state of our knowledge. Most estimates of additions to the oceans by rivers are based on total world river discharge and the average content of a given substance in the rivers. Recent estimates of the total discharge range from about 3 to 4×10^{19} g/yr. The reliability of available figures for the average concentrations of individual dissolved species ranges widely, generally with an inverse relation between concentration and probable error.

Estimates of the total particulate load of rivers range much more widely than those for dissolved load, with weight ratios of particulate to dissolved matter of from 2:1 to 10:1, although recent estimates seem to be converging toward a ratio of 3 or 4:1. Furthermore, reports on the chemical composition of the particulate load are few indeed, with no analyses available for many major rivers. Interestingly, information on the mineralogic composition of the particulate loads of rivers is much more extensive than for the total chemical composition, owing perhaps to the ease of x-ray examination of the materials. Unfortunately, the compositional variation of the phases determined by x-ray is so great that most mineralogic analyses cannot be converted safely into even approximate chemical compositions.

Obviously, international cooperation will be required to remedy the present situation. Until we know what is going into the oceans, we are helpless when asked to describe their total economies. Attempts are being made, especially through the International Hydrological Decade, to remedy the situation.

In addition to gross totals for the world, we need more specific information on the 20 or so rivers that contribute more than 90 percent of the total world discharge to the oceans. We need additional information on the composition and amount of debris contributed by ice from Antarctica and Greenland, as well as at least a total for the rest of the glaciers.

Within the oceans, some progress has been made toward the ultimate goal of understanding the distribution patterns of dissolved substances, where and in what amounts they are removed, and the biological and chemical processes involved in each case. Much recent attention has been turned toward the "budgets" of elements and their compounds.

SILICA IN SEAWATER

Here we will use silica to illustrate the kinds of studies needed, and the current status of investigation of one important chemical species.

The concentration of silica in the waters entering the oceans (13 ppm of SiO_2) is, on the average, about twice its average concentration in the oceans. The rate of addition is sufficient to renew the entire silica content of the oceans in about 20,000 years. With an ocean-stirring time of a few thousand years, the average silica molecule traverses the oceans about 7 times between entry and deposition. There is yet no agreement as to whether all the silica that leaves the oceans does so in the form of skeletal materials of organisms, or whether part is removed by chemisorption on clay minerals. Recent studies suggest that most is removed biologically. Diatoms and radiolarians deplete the silica content of surface water to almost zero and eventually die and fall toward the bottom. Re-solution while settling and after reaching the bottom destroys about 90 percent of the dead organisms. Thus, plankton production is one major control of the silica level of the deep waters.

While values are not very precise, they indicate an input to the Atlantic of about 3×10^4 g of silica each year from rivers. In addition, some 10 times the river contribution circulates to other oceans, and nearly as much re-enters from them. The net loss at the bottom is the difference between actual sedimentation (organisms that reach the sediment–water interface) and the back flux of silica from solution and subsequent diffusional and bulk flow processes.

The following is a listing of what is not known concerning this general process. Knowledge of almost every constituent is as poor or worse.

- The rate of introduction of silica into the Atlantic by rivers is probably not known within an error of ±15–20 percent.
- The percentage chemisorbed by clay minerals is unknown, but it is probably small.
- The major oceanic inflows to and outflows from the Atlantic of silica are not known, partly because of a lack of information on the rates of seawater transfer and partly because of a lack of data on silica concentrations.
- Silica additions to ocean water by volcanism and related phenomena are assessable only within an order of magnitude at best.

(Volcanism may add silica directly or by postvolcanic submarine alteration of volcanic rocks.)

- Silica addition to ocean water by submarine alteration of glacial materials cannot yet be accurately estimated. Abundant diatom production in the Antarctic may be a response to silica released by submarine alteration of comminuted silicate materials. Current thought leads to a cautious conclusion that the process is important in the total silica addition budget.
- Knowledge of the silica flux upward through the sediments depends on the results from the current deep-coring program.

CARBONATE SYSTEM

Measurements of the carbonate species dissolved in seawater have been a major part of chemical investigations since the early days of oceanography, and their use is increasing steadily. Carbonate and bicarbonate are routinely titrated as a part of the determination of the alkalinity of seawater. With the development of accurate infrared absorption methods that can determine CO_2 partial pressures to about 1 ppm and of glass electrodes for pH measurement within a hundredth of a pH unit or better, and with the current availability of apparent dissociation constants for carbonic and boric acid as functions of temperature, pressure, and chlorinity over the entire range of oceanic conditions, many new applications have been found for the measurement of CO_2 pressures, total alkalinity, and total dissolved inorganic carbon. Calculation of the total concentrations of individual dissolved carbonate species is now possible to within 5 percent or so, and this, in turn, permits improved calculations of the degree of undersaturation or oversaturation of carbonate minerals.

The major problem to be overcome in studies of the carbonate system in seawater is the lack of a way to measure or calculate confidently the concentration and/or activity of a given species, such as the carbonate ion. Perhaps progress in the near future will come mostly by obtaining convergent estimates by independent methods. "Direct" measurement of concentration, perhaps by absorption techniques of one kind or another, appears to be more remote because of the complexity of the electrolyte solution and the small concentrations to be measured (the concentration of free carbonate ions is of the order of 10^{-5} moles/kg of seawater).

Much progress has been made recently in working out complex relationships among the solid carbonate species. Almost all carbonates are precipitated by organisms; they can be classified into calcites, aragonites, and magnesium calcites (1 to 15 mole % $MgCO_3$). Biochemists are beginning to elucidate the mechanisms by which cell chemistry and organic substrates control the nature of the phases that appear. Enough data has accumulated to permit rough estimates to be made of the proportions of each material organically formed and to give a base for estimates of the degree of selective loss by solution during the depositional process. The relative solubilities of pure calcite and pure aragonite are known. It has also been established that the Mg-calcites are metastable with respect to nearly pure calcite, the degree of metastability increasing with increasing Mg content. This information has been basic to understanding the decrease in aragonite and magnesium calcite tests as a function of depth, for in the deeper parts of the ocean there is undersaturation with respect to all dissolved carbonate mineral species. It also has been applied to explaining the compensation depth—the undulating surface in the deep ocean below which no carbonate minerals are found—as the lower level of saturation with calcite.

Some work on the rates of reaction between solid carbonate species and seawater has been done; the results are interesting and perplexing. It has proven to be most difficult to precipitate carbonate minerals inorganically, even at high degrees of supersaturation. Surface phenomena, such as the inhibition of growth by Mg ions, or by organic coatings, are certainly involved, but our knowledge of these is still primitive.

Seawater is, by current calculation, supersaturated with respect to CaMg-carbonate (dolomite), as well as Mg-carbonate (magnesite). Yet neither has ever been reported as a shell material or as an inorganic precipitate from average seawater. Again, the reasons are obscure.

Several recent studies in coastal waters of high temperature and salinity have demonstrated that a variety of solid carbonate species can and do form inorganically. Such results will be increasingly useful when additional studies on reaction rates have been carried out in the laboratory.

Studies of freshwater alteration of uplifted Recent and Pleistocene marine carbonate rocks, coupled with studies of postdepositional cementation of modern reefs by carbonate minerals, are producing results that need to be explained in terms of the phases produced, their composition, and the rates at which they form.

Finally, the circulation of calcium in the oceans, from its entry via rivers to its organic depositions, remains to be worked out. Its circulation is unlike that of silica, which is everywhere undersaturated, in that the top layers of the ocean are supersaturated with three carbonate minerals, while the deeps are undersaturated with all three. Tests falling into the deeps dissolve, and their calcium eventually is transported to shallower waters, there to be precipitated.

EXCHANGE WITH BOTTOM SEDIMENTS

Our knowledge of pore-water expression as a result of compaction of sediments will be increased many-fold when the cores from the Deep Sea Drilling Program have been examined in detail and sediment age and density have been determined as a function of depth. In the deep oceans, sedimentation rates are so slow that the rate of water expression by compaction is low. In a 1,000-m core, for example, porosity may drop from 80 percent or so near the sediment surface to 15 or 20 percent at the bottom of the core, but the time interval over which this has occurred may be 30×10^6 years, so that the average upward flow is only about 10^{-3} cm^3 of pore water per cm^2 per year, or one millimeter in a century. Such a rate would be subsidiary to diffusional processes.

There is much current interest in determining concentration gradients in the seawater adjacent to the bottom, but few results have been reported. Sophisticated sampling techniques are required, and these are being developed. Enough information is finally becoming available to permit estimates of diffusional fluxes. For example, the results of analyses of pore waters for potassium suggest that the diffusional flux of potassium upward through the sediments into the ocean waters is much greater than that from rivers.

Another aspect of diffusional transport that is the object of much interest and research is the genesis of the widespread manganese nodules. In one hypothesis of origin, Mn^{+2} is released to deoxygenated pore waters by minerals and diffuses upward to the water–sediment interface, where it encounters oxygen diffusing downward, and is oxidized to Mn–O solids of higher valence.

Vertical chemical profiles, beginning some distance above the sediment interface and extending down into the sediments, are a critical need at present. These should be coupled with detailed mineralogical, chemical, physical, and paleontological studies of the sediments.

MECHANICS OF BOTTOM WATER-SEDIMENT SYSTEM

In addition to the studies of chemical gradients in pore waters, much work is necessary on pore-water circulation induced by thermal gradients. Examples of this are to be found in the Salton Sea, and especially the Red Sea, where bottom layers of very hot and very salty water occur in several deep basins, seemingly the result of extensive movement of water very deep below the seabed and its subsequent rise at a new site. Many questions are involved, such as: What are the effects of replacing warm bottom water with colder, denser water? Are the waters in the sediment turned over by density displacement?

FRESHWATER INTRUSION

A significant addition of fresh water to the oceans apparently comes from underground sources in the continental shelf regions. Major subterranean freshwater movement has been found in the U.S. Atlantic Coastal Plain sediments. Many other areas also have geologic conditions favorable for transport. Not only does this flow add water and dissolved solids to the sea, but it also causes important changes in the sediments through which it flows—sediments that may never have been above sea level. Studies are needed of the amount and composition of the water added in this way to the oceanic system, and of the interaction of the fresh water and sediments. This information can be very important in connection with radioactive waste leakage into coastal aquifers.

DISSOLUTION OF FALLING MATERIALS

The preserved sediment at the ocean bottom is a strange mixture of materials that have fallen through the water column almost untouched, those that have been almost entirely dissolved, and some that have been precipitated *in situ*. There is still little data on the solution rates of various materials as a function of the changing conditions from top to bottom of the oceans. Some work has been done on calcium carbonate in its multiple forms of occurrence, as evidenced by the concept of the compensation depth, below which carbonate minerals do not accumulate. Micropaleontologists have demonstrated the selective effects of dissolution processes on faunal and floral skeletal materials. Their work is of extreme importance in explaining the degree to which the fossil record has been warped relative to existing populations.

There remain many problems, however. For example, the ocean waters in some areas are apparently undersaturated with respect to all Ca-carbonate minerals at a few hundred meters, yet attempts to explain this in terms of solution rates have found no simple correlation between the two. In fact, the solution rate seems to increase markedly at several thousand meters. Various suggestions have been made to account for this anomaly, such as organic coatings on the solids, but the strange increase in rate in deep water is still puzzling.

GASES

Studies of oceanic gases are discussed elsewhere (p. 19). One aspect, however, is growing rapidly in interest and importance—the interactions between minerals and organic degradation products in the waters and sediments. Most sediments, even in the deep oceans, are probably oxygen-deficient below the sediment–water interface. Where organic material is abundant, as in shelf sediments and basins, large quantities of CO_2, NH_3, CH_4, H_2, H_2S, and other gases are generated. Studies are needed of these gases, of the details of their genesis as a function of the kinds of organic matter present, and of their effects on coexisting inorganic solids. Carbon dioxide, of course, will dissolve carbonate minerals, increasing the alkalinity of the water. Hydrogen sulfide can reduce and even remove Fe^{+3} from clay mineral structures. Excellent recent work has been done on the genesis of iron sulfides in sediments, but much more remains to be accomplished.

MINERAL ALTERATION AND GENESIS

An important type of study, now in its infancy, is investigation of the phase changes that take place simply because of the reaction of seawater with the great variety of bottom materials. Basic volcanic glass, so prevalent in Pacific sediments, reacts over long periods of time to form a suite of minerals. The process apparently begins by two-way diffusion between seawater and glass while the amorphous state is maintained and eventually results in crystallization. Studies of volcanic glass and of minerals will show which phases are unstable in seawater and which are metastable or stable. The changes that take place in submarine volcanics, particularly, may have been a major factor in the long-term stabilization of seawater composition.

The genesis of carbonate–fluorapatite, the mineral of marine phosphorite, has been attributed both to direct precipitation from seawater and to a replacement of skeletal carbonate. Although both mechanisms may be operating today, the relative importance of one process over the other has not been established. Controlled laboratory experiments have produced a variety of carbonate–apatites of various carbonate compositions, some similar to those found in marine deposits, through the substitution of phosphate ions for carbonate ions in calcite. Furthermore, increases in the proportion of apatite with depth in cores composed of foraminiferal oozes indicate the importance of such a process. On the other hand, thermodynamic considerations and the petrology of phosphatic nodules argue for the direct precipitation concept. Clues to which of these processes, if not both, were operating in past geological eras to produce the extensive phosphorite deposits laid down then may come from detailed mineralogical and chemical analyses of these most curious minerals.

EXPERIMENTS ON SEAWATER

Studies of seawater to determine its "tolerance limits" are also needed. How much silica can be added to seawater without causing precipitation of a silicate? How long will organic-free seawater hold a given degree of supersaturation with calcite? What controls nucleation of carbonate minerals in seawater? Sulfate minerals? Why does magnesium carbonate remain severalfold supersaturated, apparently indefinitely? Some research along these lines has been accomplished, but far more is needed. The individual species in seawater should be assessed as to the empirical limits of their solubilities in experiments other than the classic evaporation studies, so that the variables are kept out.

RADIOACTIVE GEOCHRONOLOGIES

Radioactive age determinations in the marine sedimentary column have been limited by a lack of techniques that cover more than the time period of the last half-million years. Carbon-14 has been successfully utilized for measurements from the present back to about 40 thousand years ago in both the organic and calcareous sedimentary components. Disequilibria within members of the natural uranium, thorium, and actinium series have provided geochronologies for the last several hundred thousand years. Potassium–argon ages on volcanic glasses and minerals have provided time clocks, in some tectonically active areas, for deposits laid down 20 million years ago.

However, there is a need for reliable dating of authigenic minerals in the sediments that are more than several million years old. Perhaps success will be found in the application of fission-track dating to uranium or thorium-containing phases or in the application of the potassium–argon or strontium–rubidium techniques. The need for accurate time determinations will increase as samples from deep-drilling investigations, such as JOIDES (Joint Oceanographic Institutions Deep Earth Sampling program), become available to the marine scientific community.

7

The Ocean as a Waste Receptacle and a Source of Raw Materials

THE OCEAN AS A WASTE RECEPTACLE

Important avenues of marine chemical research may be defined by the concerns expressed today on the ability of the oceans to accept the waste products of our society. As a result of an increasing world population and an increasing per capita use of energy and materials that are subsequently dispersed to the environment in degraded forms, our surroundings are being altered. The marine environment is considered the ultimate receptacle for these discharges. Their impact may be felt through alterations in the ocean's chemical composition, in the makeup of communities of organisms, and in climate.

For example, the oceans have received significant amounts of lead as a result of the discharge of lead tetraethyl degradation products from internal combustion engines to the atmosphere and their subsequent rainout to the oceans. Surface waters in coastal areas have lead concentrations up to ten times higher than the values they most probably had twenty years ago. Whether such increased levels have any effects upon the organisms living in such waters is unknown, but the question is clearly worthy of investigation. The accidental losses of petroleum from ships and drilling sites have resulted in numerous oil slicks in coastal and shipping-lane areas. Whether such layers affect the heat budget of the ocean–atmosphere system and consequently affect climate is not known.

Monitoring the oceans for those discharges of man that conceivably could result in critical problems is being considered by a number of national and international groups (see National Academy of Sciences–National Academy of Engineering, 1970). Experience with the effective management of wastes from nuclear reactors provides a pattern that may be instructive in developing controls for other potentially dangerous materials that might be discharged to the oceans. The identification of processes and substances that can deleteriously affect the marine environment should be one of the concerns of marine chemists.

THE OCEAN AS A SOURCE OF RAW MATERIALS

Although we have indicated only a minor interest in the economic recovery of raw materials from the ocean, this is an activity that must by no means be overlooked. The chemical technology is so diverse and so highly specialized and has been so adequately described in the industrial and technological literature that we believe the interests of this report can be adequately served by a brief summary.

One of the most important raw materials in the sea is water itself. Present processes of desalination, however, such as reverse osmosis and the various types of distillation and freezing, are physical rather than chemical, although some of the principles of physical chemistry are involved. Ion exchange, successful in brackish water, is not yet practical for seawater.

A large industry rests upon the recovery of salt—sodium chloride—and the manufacture of chemical products from it, principally chlorine and sodium hydroxide, both of which serve as raw materials for subsequent products. Some by-products also result.

Magnesium hydroxide, precipitated directly from seawater, is the source of many magnesium products, including the metal itself. Much of the bromine commercially available is produced from seawater that has been acified, chlorinated, and then stripped with an airstream.

The economic recovery of minerals from the sea bottom—submarine mining—is a potential industry of considerable magnitude. If and when it is developed, it will involve much chemical technology in the treatment of the ores, such as manganese nodules and phosphorite.

The products based on organic marine resources, such as fish protein concentrate and the alginates from kelp, often depend on considerable chemical technology for their production. The manufacture of other biological products, such as vitamins, toxins, and growth promoters, although undeveloped, shows promise.

The vast number of trace metals and other elements contained in seawater offers many potential opportunities for profitable recovery in the future. A laboratory method has already been developed for recovery of uranium, and the ocean's water is optimistically considered as a future source of deuterium for the field of nuclear fusion energy.

8

Analytical Methods

Analytical marine chemistry differs from the analytical chemistry of other fields in certain significant respects, among them the requirement of carrying out operations at sea and the difficult nature of the medium, seawater. The latter, because of its complexity, is a system in which one would not ordinarily choose to make analytical or physical-chemistry measurements. It is a rather concentrated solution of a relatively small number of constituents and, at the same time, a very dilute solution of a very large number of constituents—indeed, probably all the rest of the elements. The analytical problems are also complicated by the large amount of extraneous salt material, especially in the determination of the minor constituents, which exist in the range of parts per billion.

Through all of oceanography, the requirement of working at sea has restricted the development of instrumentation. This has been particularly true for chemical analysis. Some of the important laboratory operations, such as accurate weighing, cannot be carried out at sea because of the vibration and violent accelerations of the ship. Breakage is high and some conventional apparatus is very unsuitable. The advent of plasticware was a great help. T. G. ("Tommy") Thompson, a foremost figure in the field from the 1920's to the 1950's, once described the work of the oceanographic chemist as going to sea in a small boat with a handful of Nessler tubes. Indeed, this was literally true, although the young chemists of today would probably not recognize a Nessler tube.

We have come a long way. Visual colorimetry, beginning with the Nessler tube and progressing to colorimeters, which were its more sophisticated modifications, eventually gave way to photoelectric instrumentation, culminating in the modern spectrophotometer, although it was in the late 1940's that one of the latter instruments made its first extended cruise on shipboard. A research vessel of today may often carry many of the facilities of a modern laboratory, even including ultra-high-vacuum equipment. The problems of shipboard operation have largely been overcome, if not completely solved.

It is not our purpose here to describe the methods by which each and every constituent of seawater has been or can be determined analytically. There is a vast published literature dealing with such work, and a very competent review will be found in the chapter on Analytical Chemistry of Seawater by J. P. Riley, in *Chemical Oceanography* by Riley and Skirrow (1965). An excellent practical manual for technicians is that by J. D. H. Strickland and T. R. Parsons, *A Practical Handbook of Seawater Analysis* (1968). The reviews by H. Barnes, *Apparatus and Methods of Oceanography–Chemical* (1959) should also be mentioned.

We will be content here to review some of the more general considerations, particularly the ways in which the peculiarities and characteristics of the marine environment influence the problems of analytical chemistry.

The first of these problems is that of obtaining the sample for analysis. From a practical standpoint, one must consider sampling as part of the analytical process. Not only is the ocean a heterogeneous system, with solid, liquid, and gaseous phases present, but a study of the properties of the water often reveals a new scale of fine structure, so that we cannot be sure that any samples that we take are reproducible. The biological as well as the chemical distribution is "spotty," and we are continually finding that the spots are smaller and smaller. We must take this into account in considering the precision of analytical methods.

Apparatus for water sampling has developed rapidly during the last few years, and it is now possible to readily obtain samples of any size and in containers of a variety of materials. This last consideration is important. Dissolved oxygen combines rather rapidly with the metal of conventional water bottles, which are also unsuitable when working with trace metal constituents. Plastic containers are better for these purposes but must be avoided when sampling for dissolved organic constituents. The sample may remain in contact with the water bottle for a considerable time, since it may require three to four hours to bring a sample up from great depths.

For some purposes, the analysis must be made as quickly as possible on recovery of the sample. Biological processes bring about changes in the concentrations of oxygen and the micronutrients, such as phosphate, nitrate, and ammonia, as well as in pH and CO_2. Prevention of the exchange of dissolved gases with the atmosphere requires special handling and sometimes elaborate equipment. Trace metals, and even such other constituents as phosphate, adsorb on solid surfaces, either those of the container or of suspended solids. All of this makes it necessary either to work immediately on shipboard or to somehow prevent the changes just described.

The increasing desire among chemists to study the properties and behavior of substances *in situ* in the ocean is also stimulating the development of equipment that will deliver samples in a condition as nearly as possible like that in the deep sea.

Methods of preserving water for later work depend upon the work contemplated. Filtration is often a first step, although sometimes the suspended matter is later necessary. Various preservatives, such as chloroform, thymol, or mercuric chloride, have been used to stop biological changes, but these may interfere with later analysis. Present practice seems to favor deep-freezing as a method of preservation, but the conditions under which this is done are critical to the success of the process.

To appreciate the problems that face the marine chemist, one needs only to consider the wide range of the constituents of seawater that are involved. These can be grouped as major ions, dissolved gases, nutrient salts, trace elements, and dissolved organic matter. To these must be added solid materials that are part of the marine environment—and suspended particulate matter in the water. To deal analytically with such an array requires virtually the complete resources of analytical chemistry. Except for those that apply to metallic alloys, almost every analytical method is potentially of use to the marine chemist.

Gravimetric methods, which have a long history, still find use in the study of some of the major ions, while virtually the whole gamut of modern analytical chemical techniques has found application in dealing with trace elements. Some of the applicable techniques are radioactivation, isotope dilution, spectrography (including emission, mass, and gamma-ray spectrography), flame photometry, atomic absorption, spectrophotometry, fluorometry, and polarography. The several special types of chromatography (especially gas chromatography) are being widely used with success, and techniques based upon the Mössbauer effect and upon nuclear-magnetic and electron-spin resonance have found application.

Recent developments in instrumentation are well represented in shore-based laboratories of marine chemistry and even on some modern research vessels. A few of the latter are equipped with "autoanalyzers" that turn out almost simultaneous and continuous analyses of such micronutrients as phosphate, nitrate, silicate, and iron. These instruments make possible the collection of an enormous amount of data in a short time, often an important consideration in a crowded field program.

Progress is being made in the development of *in situ* instruments. Electrical conductivity and temperature are now often determined *in situ*, making possible a simultaneous record of temperature, salinity, and depth.

In situ specific ion electrodes are under development, but their development is dependent upon accurate knowledge of the effects of temperature and pressure upon electrode potentials and other parameters. An oxygen probe that measures dissolved oxygen *in situ* electrometrically to a depth of 300 m is now in use.

The objectives of analysis have changed with time. The early workers described the composition of seawater in terms of the salts that could be isolated from it. However, the relative quantities of $NaCl$, $MgSO_4$, $CaCl_2$, and other salts obtained depend upon the conditions of crystallization from saturated solutions and are no indication of the true content of the unsaturated solution.

Subsequently, the aim has been to determine the total quantity of each dissolved element. Such information is important in connection with the nutritional requirements of organisms.

An entirely new set of analytical tools came into use when attention shifted to the isotopic components, both stable and radioactive. Two recent trends have directed the choice and application of these tools.

The first is the widespread interest in the dissolved organic matter. The recent report of its carbon-14 age as 3,500 years has increased our curiosity as to its composition. Although the problem of isolating a representative sample of it has not been solved, a wide range of analytical methods–from paper chromatography to bioassay–is showing it to be a very complex mixture of many compounds.

The second trend that has influenced the use of analytical methods is the present interest in "speciation." It is not sufficient to know that an element is present in seawater; we must know its molecular or ionic state and how this compares with thermodynamic considerations.

It need scarcely be pointed out that these changing aims of analysis are not mutually exclusive. Indeed, they are all being pursued at the present time.

Our analytical activities also reflect the changing moods of society. When the ocean was being widely advertised as a future source of food, our analyses contributed to fisheries research and were applied to the problem of fertility. When the world worried about the possible pollution of the sea and its products by radioactive fallout, we sharpened our techniques for determining ^{90}Sr and ^{137}Cs. With public concern turning to other sources of pollution, our analytical aim is to determine such materials as DDT, mercury, and lead.

The field of marine chemistry affords a splendid example of the interrelationship of problems and techniques. The dominating problem of the fertility of the sea made necessary the development and improvement of spectophotometric methods for the determination of the micronutrients. The problems associated with the fallout from nuclear explosions led to extensive development of methods for isolating and assaying the radionuclides in seawater.

On the other hand, advances in radiochemical analysis, such as the extraordinary sensitivity made possible through radioactivation, have led to the solution of problems associated with trace elements that could not have been studied previously.

9

Marine Chemistry Tomorrow

In the past, marine chemists have been occupied almost exclusively in adapting the known techniques and principles of chemistry to the study of the properties and dynamic processes of the ocean. Frequently, the information obtained was of greater interest to biological or physical oceanographers than to chemists. This condition will continue, for there is no branch of oceanography that is not in some way dependent upon chemical information and the techniques of chemistry. The future, however, will see the emergence of marine chemistry, not merely as a branch of applied chemical technology, but as a recognized division of chemistry itself, affording opportunities to study chemical processes under natural conditions.

The ocean is a dynamic chemical system and can be studied as such. Questions are now arising about the input and output of this system. Much better estimates may be expected of the rate of inflow of solid material from the rivers and of the rate at which sediments are finally deposited on the sea floor. Evidence now accumulating indicates that consideration of this budget over long time periods will lead to better knowledge of conditions in the ocean and on the earth in the past, and possibly to some insights into the future.

There are too many unsolved problems in marine chemistry—as

indicated in the preceding chapters—to permit individual attention to them here. Progress will doubtless be made on many of these, such as that of the constancy of ionic composition and its bearing on the problem of "salinity." The chemical composition of the ocean is enormously complex, including unknown thousands of compounds, some in extreme dilution. It is impossible to predict which of these will prove to be of future interest and importance. There are, however, a few more or less obvious directions in which interest is turning.

The organic chemistry of the sea is a field that has been late in developing but in which we may expect steady advance. The study of natural organic products from the sea has scarcely begun, but it will develop, perhaps with the emergence of a "marine pharmacology."

The "organic burden" of seawater, amounting to not more than a very few milligrams per liter, may be the source of startling discoveries. A study of its composition can possibly settle some of the unanswered questions about past and present processes in the ocean. It is a source of tracers for the links of the food web and for the various events in the cycle of decomposition and regeneration. It contains the substances responsible for organic chemotaxis.

As it has in the past, the advance of marine chemistry in all its branches will continue to depend upon the development of analytical methods suitable for the marine environment. One of the greatest existing needs that the future may be expected to fill is that of *in situ* methods applicable at sea—methods for chemical measurement without bringing the sample out of the water. This will involve electrochemical applications, particularly the development of specific-ion electrodes.

There is a need for the development of methods of remote chemical analysis from the air or from space. Optical methods seem promising for this purpose. At the other extreme, there is a call for the greater participation of chemists in man-in-the-sea programs, which will involve consideration of the effects of elevated pressure on laboratory procedures.

The use of sophisticated instrumentation, such as scanning electron microscopy, nuclear magnetic resonance, and Mössbauer and laser Raman effects, will expand in both seagoing and land-based marine chemical laboratories.

The rapid advance of marine science involves the participation of more and more people who are making more and more measurements. This situation requires the development of better methods for managing the increased quantity and the quality of the data. For the former—the recording, storage, and digesting of the data—computer techniques are available and are becoming common.

Quality control, however, needs more attention. For example, it has been reported that much of the chemical data produced by the International Indian Ocean Expedition is unusable because of doubts about its accuracy. Such reports are a perennial source of confusion in marine chemistry. Better calibration, universal standards, and interlaboratory comparison are essential if we are to continue our present field methods, in which independent investigators make measurements that are presumably comparable.

Intercalibration projects, in which several investigators carry out analyses simultaneously on the same samples under field conditions, are an obvious means of attacking this problem and have been carried out with varying success. This effort should be continued. Another possibility that deserves trial is the establishment of several "standard stations" at convenient locations that can be visited periodically by expeditions and at which chemical analyses can be made at standard depths for intercomparison. Real variations of composition will undoubtedly take place with time, but it is not likely that they would be of such a magnitude as to prevent the comparison of analytical methods.

A field that has been mostly overlooked by marine chemists is that of surface phenomena. However, the future will see a number of developments in this direction. These will probably concern, first, the invasion–evasion kinetics of gas transfer and the various effects of surface films at the air–sea interface. The "slicks" resulting from oil wells and damaged tankers excite much attention, but there are also important effects from less obvious films (even to the monomolecular extreme) that have largely escaped notice but that influence our weather.

The various solid–liquid interfaces are also important as the probable loci of many chemical processes. This is an almost untouched field that deserves and will get more attention in the future.

Marine chemists will also give closer attention to the sediments and to their interactions with seawater. The Deep Sea Drilling Project is making available new and important information about the composition of the interstitial water within the sediments—information that has a bearing on many marine problems. A whole new field of research has been opened up by the discovery of concentration gradients within the interstitial water. Such gradients could exist only as a result of an exchange between the interstitial water and the supernatant water column. There has been found to be a close parallel between conditions at the air–sea interface and those at the water–solid interface, although the respective time scales make the kinetics vastly different.

Our data on the composition of the ocean and its variation in space and time has been collected piecemeal. Chemical analyses have generally been made incidentally during other field operations, but there has been little effort to obtain synoptic chemical measurements that could lead to really broad generalizations. Combining data from independent operations is of dubious value until the problem of intercalibration is solved. The Geochemical Ocean Sections Study project (GEOSECS), now just past the planning stage, is a good example of what is needed in this regard. GEOSECS entails the detailed measurement of a number of oceanic chemical parameters along several Arctic-to-Antarctic sections at all depths. It will provide, for the first time, a comprehensive set of physical and chemical measurements that have been made on the same water samples. This project may well furnish us with the data necessary for the solution of many marine chemistry problems.

It is particularly encouraging that this project has been given adequate financial support. We hope that its success is an indication of the future. In the past, chemical oceanography has been at a disadvantage in the competition for funds, except as it supported the apparently more "practical" areas of physical, biological, and geological oceanography.

In the immediate future there will be great concern over the pollution of the environment. Since the ocean is an obvious and inevitable reservoir for the final deposition of waste materials, we must have better information about the fate of the various present and potential pollutants in the marine environment. The further development of nuclear energy will aggravate many of the problems we already have and perhaps present us with new ones. The introduction of radioactive wastes to the marine environment is a special case discussed in greater detail in another publication of the National Academy of Sciences (1971).

In the economic field we may expect a steady advance in the recovery of useful chemical products from the sea, but, with the possible exception of pharmaceuticals, there seems little prospect of startling new discoveries.

The desalination of seawater will become cheaper and more common. It remains to be seen whether other techniques, such as reverse osmosis, will displace or supplement the now well-established distillation methods.

The recovery of new chemical products from the sea, such as trace metals, depends upon the solution of engineering problems. In many cases, to solve these cases will require better knowledge of the speciation or actual ionic state of the elements concerned.

Such knowledge is also required for better understanding of the physical chemistry of seawater. All the newer methods of chemical investigation will be brought to bear on this problem.

It has already been mentioned that a new class of workers is coming into the field—young students interested in studying chemistry in the natural marine environment. Like the other branches of oceanography, marine chemistry needs both generalists and specialists; it will continue to be concerned with the general problems of the ocean and can use the breadth of view of those whose training has involved broad understanding of these problems. As in the past, chemistry will make some of its most important contributions to oceanography in cooperation with the other disciplines.

The rapid advance of marine chemistry needs new ideas and the latest chemical techniques. The most important contributions to this advance may well be made by those who have come up through conventional chemical training before turning to marine problems.

Training may therefore be along either of two paths: through oceanography departments of institutions where graduate work in general oceanography is combined with specialization in marine chemistry; or through the conventional course leading to a graduate degree in chemistry. In the latter case, some familiarity with marine work may be obtained either concurrently or subsequently in a postdoctoral period.

The number of students already on these two educational paths would seem to be sufficient to satisfy present needs for manpower in the near future.

References

Barnes, H., 1959. Apparatus and methods of oceanography–chemical. Interscience, New York, 341 p.

Harvey, H. W., 1956. The chemistry and fertility of seawater. Cambridge University Press, London. 224 p.

Journal of Marine Research, 1958. T. G. Thompson seventieth anniversary volume. Volume 17. Sears Foundation for Marine Research, New Haven, Connecticut. 583 p.

National Academy of Sciences–National Academy of Engineering, 1970. Wastes management concepts for the coastal zone: Requirements for research and investigation. Washington, D.C. 126 p.

National Academy of Sciences, 1971. Radioactivity in the marine environment. Washington, D.C. 272 p.

Revelle, R., and H. E. Suess, 1957. Carbon dioxide exchange between atmosphere and ocean and the question of an increase in atmospheric carbon dioxide in the past decades. Tellus 9, 18.

Riley, J. P., and G. Skirrow [Ed.], 1965. Chemical oceanography, Vol. 1 and 2. Academic Press, London. 712 p. and 508 p.

Strickland, J. D. H., and T. R. Parsons, 1968. A practical handbook of seawater analysis. Fisheries Research Board of Canada Bulletin 167. Ottawa. 311 p.

A Bibliography of Recent Literature

American Chemical Society, 1964. Chemistry and the oceans. Chem. Eng. News Spec. Rep. 48 p.

Hill, M. N. [Ed.], 1963. The sea, Vol. II, p. 1–231. Interscience Publishers, New York.

Hood, D. W., 1966. Seawater chemistry, p. 792–799. *In* Rhodes Fairbridge [Ed.], The encyclopedia of oceanography. Reinhold Publishing Corporation, New York.

Horne, R. A., 1969. Marine chemistry; the structure of water and the chemistry of the hydrosphere. Wiley-Interscience, New York. 568 p.

Horne, R. A., 1970. Sea water, p. 107–140. *In* Chow Ven Te [Ed.], Advances in hydroscience, Vol. 6. Academic Press, New York.

Lange, R. [Ed.], 1969. Chemical oceanography; an introduction. Universitetsforlaget, Oslo. 152 p.

Martin, D. F., 1968. Marine chemistry, Vol. 1, Analytical methods. Marcel Dekker, Inc., New York. 280 p.

Martin, D. F., 1970. Marine chemistry, Vol. 2, Theory and applications. Marcel Dekker, Inc., New York. 451 p.

Morowitz, H., 1968. Energy flow in biology. Academic Press, New York. 179 p.

Richards, F. A., 1966. Chemical oceanography, general, p. 186–191. *In* Rhodes Fairbridge [Ed.], The encyclopedia of oceanography. Reinhold Publishing Corporation, New York.

Riley, J. P., and R. Chester, 1971. Introduction to marine chemistry. Academic Press, London. 466 p.

Stumm, W., and J. J. Morgan, 1970. Aquatic chemistry; an introduction emphasizing chemical equilibria in natural waters. Wiley-Interscience, New York. 583 p.

Vinogradov, A. P., 1953. The elementary chemical composition of marine organisms. Sears Foundation for Marine Research. New Haven, Conn. 647 p.